H. O. Curling

Hints on the Use and Handling of Firearms

Generally, and the Revolver in Particular

H. O. Curling

Hints on the Use and Handling of Firearms
Generally, and the Revolver in Particular

ISBN/EAN: 9783337250935

Printed in Europe, USA, Canada, Australia, Japan

Cover: Foto ©berggeist007 / pixelio.de

More available books at **www.hansebooks.com**

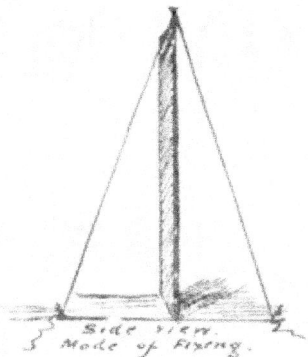

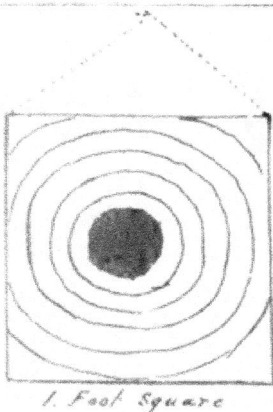

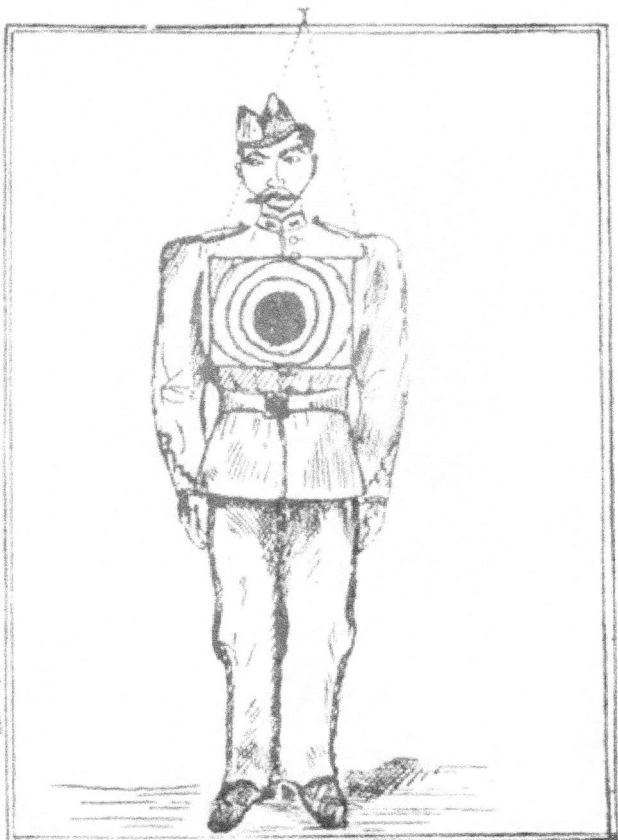

Side view.
Mode of Fixing.

1 Foot Square

2 Feet

HINTS ON THE USE

AND

HANDLING OF FIREARMS

GENERALLY, AND

THE REVOLVER

IN PARTICULAR.

BY

LIEUT. H. ONSLOW CURLING,

C. L. A. B.

'*Nunquam non paratus.*'

LONDON:

DULAU & CO., 37 SOHO SQUARE.

———

1885.

LONDON:

PRINTED BY STRANGEWAYS AND SONS,
Tower Street, Upper St. Martin's Lane.

HINTS ON THE USE AND HANDLING OF FIREARMS,

&c. &c.

'He, that rides at high speed, and with
His pistol kills a sparrow flying.'

SHAKESPEARE : *Henry IV.*

THE National Rifle Association may fairly claim the honour of introducing, at their meeting in July 1885, the subject of Military Revolver practice in this country. For years past the want of such a movement has been felt, but the many obstacles to be overcome have been so vast that no one seems to have cared to venture upon the matter, and so it has slept.

The great drawback has been, and is now, to find suitable ranges anywhere near London. Such ranges, the use of which is enjoyed by our Citizen Army, are insufficient, and the expense of keeping them up is considerable, falling heavily upon the corps to whom they belong.

The National Rifle Association, although they offered some 40*l.* in prizes, and provided not only revolvers but ammunition, for a small consideration, or entrance fee, met with but poor support ; but it should be borne in mind that this was the first year of such a competition, and it was in consequence not generally known of. Very little was known of the movement till it actually took place, and then only when noticed by the press the day after its introduction.

Again, it should be remembered that the entries were restricted to officers, warrant officers, and petty officers, of her Majesty's land and sea forces, and doubtless this restriction accounted for the spare attendance. Every Englishman belonging to the auxiliary forces should hail with pleasure the opportunity offered of making himself master of this useful weapon ; one that in skilled hands is most deadly at long or short ranges, and a thorough knowledge of the use of which might at any moment be the means of saving another's life from an opposing force when no other weapon was at hand.

The difficulty in using even an ordinary pistol with accuracy is, and always has been, an acknowledged fact, as it requires great practice to enable a man to make his mark as a crack shot. Some men would perhaps miss a haystack at twenty yards, while others, with little practice, soon become excellent shots at very small objects. It is marvellous the accuracy with which the professional burglar has of late years used his revolver against the police and others ; but it may be accounted for by the fact that these men use a small, light weapon, easily carried and much easier wielded than the military regulation revolver, which weighs 2 lbs. 8 oz. ; that they invariably take what may be termed flying shots—and it should be remembered that a full-sized man at comparatively close quarters presents a very large target. I venture to affirm that if these burglarious minions of the moon, who make night hideous, were compelled to stand before a Martini-Smith target (a foot square) at twenty yards, with a military regulation revolver, they would make but sorry marksmen.

The use of the military revolver is acknowledged to be a question of great importance, as one not only affecting those who embrace the profession of arms, but those who travel; and as no one knows when he may be called upon, or where he may be, it is imperative that he should gain a thorough knowledge of every minor detail, most useful in the hour of need, and which will enable him not only to protect himself with confidence, but to come to the assistance of the weak should occasion require.

It is to be deplored that what once formed part of the education of a gentleman—*i. e.* the use of the small sword and broadsword—should have been so thoroughly neglected of late years in this country. That part of the education of youth seems to have become quite a secondary consideration.

General Sir Charles Napier has truly said, ' Young men have all the temptations in the world to pleasure, none to study; consequently, they some day find themselves conspicuous for want of knowledge, not of talent.'

The introduction of the Breechloader has revolutionised firearms. When we look back upon the extraordinary achievements of arms during the present century, with the comparatively crude weapons then in use as compared with the marvellous inventions of the present moment, it is simply astounding what results were obtained.

The terrible work done by the old Brown Bess, with its unique flint-and-steel lock of its day, at Waterloo and elsewhere, is now matter of history. In those days artillery and cavalry had a chance of existence in the field, they have scarcely any now. The old flint lock, although it has had its day, has done its work well, and is entitled to veneration. Many a noble fellow has bit the dust from its spark, and England's first and greatest battles were fought and won by its aid. The Nipple and Percussion Cap came next into use, and subsequently the Breechloader; but since Rifles have superseded military smooth‑bore weapons, the old spherical ball has been condemned.

The breech‑loading rifled arm of the present day may be looked upon as a marvel of modern ingenuity; as combining exquisite manufacture, extraordinary precision, and un‑ equalled range. The latter may be accounted for by the conical shape of the bullet, and the rotary motion given thereto by the grooving of the barrel; and lastly, from the full force of the evolution of gas consequent upon the powder being enclosed in a copper tube which is inserted in the breech when loading the piece.

The barrel of the Breech‑loading Rifle is by its own action of firing kept comparatively clean, as compared with the old Muzzle‑ loader; for with the breech‑loader any fouling of the barrel is driven out by the discharge, and the powder in the cartridge is kept per‑ fectly free from any contamination with the moisture adhering to the barrel by its copper case and being inserted in the breech; whereas in the old muzzle‑loading weapon the barrel, after the first discharge, becomes lubricated, and consequently a portion of the powder poured down the barrel adhered to its moist sides, thereby becoming deteriorated and decreasing the explosive force. As a weapon of precision the Snider is perhaps preferable to the Martini-Henry ; but, of course, this is matter of opinion.

The sportsman of the good old school would be somewhat astonished, and would perhaps feel uncomfortable, upon finding himself armed with a breech-loading fowling-piece of the present day, particularly as prejudices are strong and obstinacy very prevalent among some people, and the keen eye of the old sportsman would view the modern innovation upon his rights—as he would probably call them—with dread, suspicion, and distrust.

It is a fact, even at the present time, that there are many old farmers in England who use their ancient flint-and-steel fowling-pieces from choice in preference to modern weapons.

The cool old sportsman of days gone by would sally forth in quest of game, having

previously overhauled his lock, and, if necessary, adjusted a new flint, with as much care
as an angler would examine his tackle pre-
vious to a day's sport, as he well knew that
success depended upon vigilance and care.
There was no blustering and banging away
in those days, as soon as a bird rose, as is
unhappily too often the case now-a-days,
resulting in either blowing the bird all to
pieces or probably missing it altogether.
No, the keen eye of the old school would
coolly watch his bird rise, take a pinch of
snuff, cock his piece, cover his bird, and
then bring it down, allowing it to get well
away before drawing the trigger.

Many a young gentleman calling himself a
sportsman knows little of the capabilities of
the weapon he wields, and cares less; his
whole aim is to see how many head of
game he can bag, and to blaze away is
the order of the day, to the astonishment
of poor Ponto, who, if he chance to run
within range, sometimes gets a charge of
shot in his tail.

In the Royal Navy the use and practice of

the pistol, and latterly of the revolver, has always been kept up. Consequently the Jack Tar knows more about the pistol and the military revolver than most men give him credit for. In boarding vessels, for instance, the pistol was one of the arms used. The importance of the revolver movement as inaugurated by the National Rifle Association has resulted in the formation of a club called 'The Metropolitan Revolver Club.' This Club, which is in its infancy, has many obstacles to surmount, but it is to be hoped that the Provisional Committee will be able to carry out the object in view, which is, according to the programme, as follows :—

'That this Club be formed, having for its object the provision of facilities for acquiring a thorough knowledge of and proficiency in the use of the Military Revolver.'

DUDLEY WILSON, Esq., 2 Pall Mall, is the Honorary Secretary, and may success attend him.

To the inexperienced, the revolver is,

perhaps, as deadly a weapon as can well be handled ; and to no class is this fact so well known as to naval and military men. The many deplorable accidents resulting from the incautious handling of firearms is terrible to contemplate ; and sportsmen and military men have frequently fallen victims to carelessness, to say nothing of novices. The unfortunate part is, that foolish and inexperienced people often inflict misery upon innocent persons ; unintentionally, it is true : but they are none the less guilty. Firearms should be looked upon as a kind of machinery, which no one in his senses would attempt to handle unless he knew the use of them.

The abominable practice of those to whom firearms belong, or those in the charge or care thereof, of keeping or leaving such weapons loaded, so that they may at any moment fall into the hands of children, or perhaps, what is worse still, inexperienced adults, is most seriously to be condemned, and may be designated really as a criminal act, which ought to be summarily punished.

It is an act which has no real motive, no real *bonâ fide* object, and is lawless and idle in the extreme,—an act which has resulted in the death of its thousands, and the maiming of even more.

A weapon should never be brought within the portals of a man's house loaded; the breech-loading cartridge can be easily withdrawn. If the piece is a muzzle-loader it should be discharged after the day's sport is over; ammunition is really not so very costly as to require to be husbanded at the probable cost of a serious accident, or perhaps a fellow-creature's life. This rule cannot be too strictly adhered to. Some years ago it was my lot to be staying with a gentleman of eccentric habits, a man of violent temper, and when in one of these fits really not answerable for his actions. I was aware that he kept a full-sized revolver loaded with ball, and capped, in his dressing-room. I confess I was coward enough to let this matter trouble me. I felt I could stand up and face death with any one in the field, fighting in a good cause and armed as others; but to be taken advantage of at any moment, and perhaps shot down like a dog, was rather too much. I therefore resolved in my own mind, not only to disarm my friend but to render his weapon useless ; but how to accomplish this was the question, as to raise any suspicion would perhaps bring down wrath upon my own head. I therefore resolved to leave everything precisely intact till an opportunity should present itself. The very next day the time arrived, and during this Grand Turk's absence I hastily removed the caps from off the nipples of the revolver, and having exploded them upon the nipples of his double-barrelled gun, I pinched them back into their original shape and replaced them on the revolver. I then put the box of caps into my pocket and felt perfectly secure, and could have sat and been fired at without the slightest fear. This gentleman shortly afterwards was seized with paralysis of the brain, and ended his days in a madhouse. No one, I believe, ever suffered any inconvenience from the revolver, and what became of it I know not.

If leaving weapons about is necessary

(which I do not for a moment admit), then most assuredly they should be **rendered** harmless by being left unloaded, and thus the means **of** rendering **them destructive** would be **kept out** of the **way of meddlers.** All ammunition **should, as a rule, be kept in** some secret and safe place, **and** always **under** lock and **key.** Every **man knows that edged tools** are dangerous, consequently that **the** leaving loaded **firearms** within **the reach of** anybody who **may chance** to **come across** them is simply leaving **means of** destruction unprotected, and **he** should **bear** in mind that this mischief **of his own neglect** might accidentally **at** any moment be wielded against himself.

> ' How oft the sight **of means to** do ill deeds,
> **Makes** deeds ill done.'
> SHAKESPEARE : *King John.*

The responsibility of those possessing firearms is **great, and** proper precautions and proper care cannot be too strictly enforced. Care costs nothing, and may be the means of preventing **loss of life and** many a deplorable **accident.** The precautions necessary to be borne in mind in the safe use of firearms for one's own protection, as well as the protection of others, are voluminous, and so varied are they that it is with difficulty they can be all dealt with in this little treatise; it is only therefore proposed to mention some of them, and detail a few important hints for the guidance of the unwary. Generally speaking, if a man will not exercise a little gumption, care, and discretion, when in the society of a shooting-party similarly armed as he is himself, he must put up with the consequences. Accidents in properly regulated families should never happen. Since the introduction of the breech-loader there is no excuse for any man carrying a loaded weapon and swinging the muzzle of it about when carrying it on his shoulder (which is often done), bringing every one in his rear in the line of fire of the piece. A man can load his piece now when he arrives upon the ground in a moment; and should a bird rise, with the present facilities given by the breech-loader, there is ample time to load and bring the bird down without the slightest difficulty.

For any man therefore, when not in the field, to strut about with a loaded weapon in his possession now-a-days is simply bombastic tomfoolery.

To carry a gun gracefully and properly is an art. It should never be so carried or wielded as to be a risk to the possessor, or any one. The following are a few ways how a gun should be carried :—For safety, when commencing sport, the right hand grasping the piece at the small of the butt, the butt resting on the right hip or thigh, muzzle up. The weapon can then, on the rising of game, be at once safely presented.

When carried on the shoulder it should be always with *lock down :* this mode will so elevate the barrels that the muzzles are far above the heads of any one ; even when at close quarters, on the march, or when approaching or returning from cover, this way will be found easiest and with the least possible fatigue, as the weight of the weapon is centered in the stock held in the right hand. To relieve the shoulder pass the

B

hand up to the small, or neck of the butt ; at the same time seize the butt with the left hand, then raise your gun to a perpendicular position, carry it across the body, and place it on the left shoulder. The left shoulder can be relieved in a similar manner, *i.e.*, pass the left hand to the small or neck of the butt, at the same time seize the butt with the right hand, raise the gun to a perpendicular position, and carry it across the body and place it on the right shoulder. Never present, much less fire, when any person, whether keeper or beater, intervenes or is near the bird. Never fire over any one, even if he what is called 'ducks,' or stoops to allow of your doing so. A keeper or beater should never be encouraged in, or allowed to 'duck' or stoop; the practice is a bad one, and should be for ever discountenanced. If no one fired over a ducked body the habit would soon fall into disuse. Sportsmen and others would do well to bear in mind that an accident deprives the injured man from earning his livelihood, and the poor wife and children suffer: better to forego taking a shot for safety sake and let the bird escape for another day than run any risk. This should be made a hard-and-fast rule among sportsmen, and a law of sport.

The left hand should never be placed upon the gun till the bird has risen and *all is clear* ahead. Coolness in the field is everything; there should be no blundering, no hurry; a man who knows the capabilities of his gun can afford to be cool. He knows but too well there is no occasion for haste; the cool hand would pause after the bird rose, and give it time to get fairly away before presenting.

A gun should never be so wielded as to bring its barrels in line with any one, or the barrels athwart any one. When quite a youth I remember being in the field, when one of the party becoming fatigued from the effects of a tight boot handed me his gun; the friend, who evidently did not appreciate the confidence placed in the youngster, kept aloof—well to the right; presently a bird rose, I hesitated; looking at the bird. 'Fire! Fire! why don't you fire, sir ?' exclaimed the old gentleman with some warmth. 'How can I,' cried I, 'with those peasants at work right in front ?' The effect was marvellous. The old gentleman, thoroughly appreciating the caution, at once joined me, and I had the benefit of my full share of the sport.

Firing when in thick cover and from behind hedges should be conducted with caution, and with a knowledge that all is clear on the other side.

Little observation will show whether your companion has been accustomed to the use of firearms. A man of reckless temperament, one who would blaze away blindly, a devil-may-care sort of fellow, should be avoided ; give him a very wide berth, and keep the gentleman well on your extreme left. If you can shunt him altogether so much the better. A gun should never be carried in the field at the trail ; should never be carried under the arm, hugging the lock ; should never be carried muzzle down, so that by an accidental slip, or stumble, or fall, the barrels may become choked with earth (which would burst the muzzle if not removed before firing) ; should **never be carried** transversely **across the body** with barrels **pointing left.** When **shooting, a** man **should be as much upon his etiquette as he would be in my lady's** drawing-room ; should mind his P's and Q's, and remember **that when in a china-**shop he should refrain **from carrying his** umbrella **under his arm.**

As a fact, the closing of one eye in taking aim is unnecessary. The **complete angle of sight** upon a **given** object **can only be** obtained by **the use of both eyes.** Con-sequently two **objects** cannot **be** seen dis-tinctly or clearly at **the** same instant, one is clear while the others are blurred or misty ; **hence it** stands to reason, that in laying a gun **the top** of the notch **of the** hindsight, the **apex of the** foresight, **and** the object, can **be** brought **into line** as accurately **with both** eyes open **as with one** closed.

An **artilleryman can lay a** gun perfectly **without closing** one eye. **The eyes should not be less** than 12 inches from **the hind-**sight, if from 2 to 3 feet so **much the better,** and a more accurate aim will be the result.

Upon the principle that the hand follows the eye, a sportsman fixing both eyes upon his bird can take as perfect an aim as he could with one eye closed.

This rule applies equally to all arms.

A man when in the field or at practice should keep his eyes about him ; he should remember whom he is with ; that he may be covered by a friend's gun or rifle at any moment, and that as the abominable and unnecessary proceeding of carrying weapons loaded, when not actually in the field, is the rule rather than the exception, he may perhaps find himself accidentally pinked at any moment, and when he little expects it.

I remember some years ago the magnificent solemnity of a military funeral was brought to a somewhat ludicrous termination by one of the firing party shooting his comrade in the stern. How the accident really occurred I never could learn ; but it was a fact that the rear-rank man managed somehow to discharge his rifle, and pretty nearly blow off the tail of his comrade's tunic.

The wounded man, who was more frightened than hurt, seemed not at all to relish the joke. An old lady came to the rescue.

This good old soul seems to have been in the habit of carrying a flask, and, graciously offering the ' pocket pistol,' suggested a drop of the creature. The offer was most readily accepted, but, I regret to say, the terror of the injured man was so great that he emptied the flask. He had evidently had enough of soldiering and ' villainous salt-petre,' for the very next day he sent in his resignation.

At ball practice men should refrain from talking, joking, and that ungentlemanly pas-time known as *horse-play*. Their attention should be directed to what they are about to do and what others are doing, and they should leave frivolities for some other time.

Many accidents in the field have occurred when getting over stiles, gates, hurdles, stone walls, and even through hedges.

Within the beautiful glades of Kensington Gardens stands a lasting memorial.

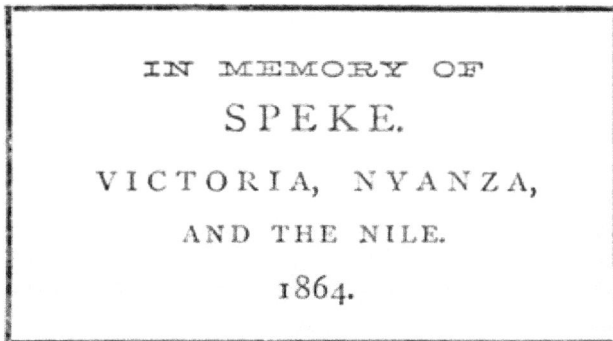

> IN MEMORY OF
>
> # SPEKE.
>
> VICTORIA, NYANZA,
>
> AND THE NILE.
>
> 1864.

Here is a terrible record of an awful death through carelessness. A noble life lost, sacrificed in a moment. Poor Speke, who had faced death often in many forms, met it at last by his own hand.

While out shooting, in getting through a hedge he dragged his fowling-piece after him, the muzzle towards his own body, when, the lock becoming entangled in the brambles, his immediate death was the result. Such a piece of foolhardiness on the part of a man accustomed to the use of firearms is astounding.

Use dulls the edge of caution, and some men, unhappily, who are accustomed to deal constantly with weapons and ingredients of destruction, become not only careless but indifferent and callous.

There is a class of men who, if not kept under surveillance, would probably be found smoking their pipes in a powder-magazine, or while sitting upon a barrel of gunpowder.

Men are too prone to carry their weapons at full-cock. This should never be done. If alone, when getting through a hedge or over any *impedimenta* the weapon should be laid on the ground, parallel with the hedge, if possible. After getting upon the other side, the weapon should be drawn through with the butt end towards the person.

If you have a comrade or keeper with you, hand him the weapon, muzzle up; get through yourself, and then take the weapons from him, *muzzle up*, and he can follow you with safety. Always place your weapon upon half-cock (it should never be at full-cock) before attempting to go through a hedge or over a stile.

When two or more gentlemen take the field together, it is advantageous to work the ground in the formation of échelon.

The whole field will by this means be thoroughly searched for game, and each man can fire clear of the other, commanding his own ground and the whole field within the range of the respective guns.

When about to commence practice with the rifle or revolver the firing party should be placed well to the front, and should never load, or be allowed to load, until all preliminaries are arranged, and the words, 'Ready! go on!' are given.

This command or caution will, of necessity, place every one upon his guard.

When the piece is loaded, the finger with which the trigger is drawn should on no account be placed within the trigger-guard till the weapon is raised and the aim about to be taken; and with the rifle until the weapon is presented, after being put upon full-cock.

In firing with a pistol, or revolver, the proper finger with which to draw the trigger is the second finger, not the index finger, as generally used. The index finger should be placed horizontally along the barrel, on the side of the weapon, which is most important

—which, as a means of securing **steadiness**
and leverage, tends not only to reduce the
difficulty of the pull, but also tends **to prevent depression** of the muzzle, which is sure
to take place if the forefinger **is used,** particularly when the **trigger has the minimum**
five-pounds' **pull.**

When a gun, **rifle, pistol, or revolver, is at**
full-cock, and it is desired **to place it upon**
half-cock, as is often done, it should **be**
so altered, with great care, **as follows :—**

The **hammer** should **be lowered gently to**
the full extent of the spring, and should then
be carefully drawn back till the distinct *click*
of the half-cock is heard ; **then the** weapon is
as safe as an arm can be when loaded, and
cannot be accidentally discharged.

To place a weapon from full to half-cock,
by not lowering **the** hammer to the full extent of the spring, and then drawing it back
to half-cock **as before** described, is **a most**
dangerous practice, **as the** hammer may not
be properly inserted in **the** clip, and an accident might be the result. **A** man once having
taken up his position at the firing-point, and

having loaded his piece, should never return into the company of his comrades till his piece (particularly if a pistol or revolver) is discharged, or till all its chambers have been expended. If it is necessary for him to rejoin his comrades after his piece is loaded, or after any of the chambers have been expended, he should leave the weapon behind him at the firing-point, and should place it, *muzzle down*, in a hole or slot purposely made in the table before him to receive it, which hole in the table should have the word 'LOADED' written legibly near it.

If there is no table, then the weapon should, if at full-cock, be placed upon half-cock, as before described, and then laid carefully upon the ground, muzzle pointing towards the target, and slightly inclined to the left thereof, so as to be clear of it, which will allow of the target being examined, if necessary, without the examiner coming within the direct line of fire of the weapon ; but the table with a hole in it is the safest method, and is recommended.

A couple of stakes with a rope from the

firing-point to the target should be used, as a precaution to keep back idle curiosity-seekers from placing themselves within danger on the firing party's left.

No one should, upon any pretence whatever, place himself, or be allowed to place himself, on, or even near, the firing party's left side. The reason is obvious, as it will be found invariably in practice that a man, when loading with a breech-loader, will naturally incline the muzzle of his piece, and so innocently place those immediately upon his left within its range.

If it is necessary to address a man when at the firing-point all interlocution should be addressed to him on his right; so the Instructor should place himself on the right and rather behind the practitioner, and as close to him as convenient, so as not to incommode his freedom.

Some men are naturally nervous, particularly when at ball practice, and for this reason all but novices should be left alone, as they will perhaps make better scoring if not interfered with.

All spectators should take ground well in rear of the alignment of the firing-point, and on its right flank. The practice of taking up weapons and going through the pantomime of pointing them at the target, or pointing a weapon at anything when not at actual practice, is idle, and is to be condemned.

Weapons set aside for practice should never be meddled with.

The party who takes his turn (if firing with revolvers) should receive his weapon unloaded, *muzzle up,* with the necessary amount of ammunition, from the Instructor or Superintendent in charge ; he should then step to the front or firing-point, load his piece himself, and get rid of his cartridges as quickly as a due regard to careful aim, &c., will admit ; then return his piece, *muzzle up,* to the Instructor, who will carefully examine it and satisfy himself that all the chambers have been expended.

Should a revolver miss fire, it is most important that great caution should be used, as it will sometimes '*hang fire,*' which the cartridges of all weapons are liable to do at times.* When a cartridge does not explode the revolver should be held in the same position as much as possible, muzzle to the front, or downwards, for a few seconds; should it not then explode it may be examined, the non-exploded cartridge removed and condemned, and a new cartridge put in its place. On no account should the condemned cartridge be placed with or near live cartridges.

Firearms should never, under any pretence, be pointed at anybody; even if unloaded, such a practice is foolish and unpardonable. No soldier except in action would ever think of doing so, and no gentleman could.

The thoughtless practice of relinquishing one's weapon into the hands of a friend, or, even worse, a stranger, is against all military rules, and in any case is strongly to be condemned, and no excuse will palliate such an offence; not even the assurance that the piece

* I have known instances of pistols and fowling-pieces hanging fire for two or three seconds after the hammer has fallen, and then suddenly go off.

is unloaded. A brother-comrade in the same regiment is, perhaps, the only exception; but even this is objectionable, except in extreme cases. As a rule, a soldier should never *relinquish* his piece, even to a General or a Field Officer.

Firearms generally, and particularly revolvers, when loaded or unloaded, should never be laid upon a table so that the muzzle can accidentally cover any one. If they must be relinquished by the owner they should be placed in a corner of the room farthest from the door, leaning against the wall, muzzle down, so that they cannot fall. If loaded they may, when practicable, be laid upon a side-table, muzzle towards the wall. Guns or rifles should be stood muzzle up in their place in the rack, or, if there is no rack, then in a corner of the room farthest from the door, to prevent surprise. No weapon of any kind should be carried or put down, or left at full-cock, and no loaded weapon should be left unprotected. They should, if loaded, be in the charge of some trustworthy and responsible person; but in the time of war no man would be so foolish as to relinquish his piece, either by night or by day.

To sportsmen and others, with the great facilities for loading and unloading afforded by the breech-loading system, there can be no excuse for leaving a weapon charged when it can so easily be rendered harmless.

There are many theories as to the proper way to present a pistol or revolver.

Every man has some idea upon the subject, and perhaps it would be well to leave every one to his own devices ; but at the same time a suggestion here, as we are upon the subject, may not be out of place.

The French carry the weapon muzzle up, the lock of the piece in line with the ear. Upon taking aim, the muzzle is gradually depressed till the object it is desired to hit is covered. This is no doubt a very good way ; but when firing at any distance beyond a point-blank range it necessitates, firstly, the depression of the muzzle to cover the object, and secondly, the necessary elevation must be taken so that the ball may be carried the required distance, and so *hit* the object.

This position of holding the weapon when at practice commends itself on the ground of safety.

The preferable way, perhaps, is the old duelling style ; that is, to hold the weapon muzzle down at the full extent of the right arm, standing sideways or three-quarters left, showing as small a front as possible, the eye to be fixed steadily upon the bull's eye or centre of the target or object, then gradually raising the arm to the required elevation. Should the distance be beyond the point-blank range, after covering the bull's eye continue to elevate till the required elevation is reached : by then steadily and firmly increasing the pressure of the second finger on the trigger the desired result will be obtained. Suddenly drawing or jerking the trigger should be avoided.

By the latter means the object is covered at the same time as the foot of the target is covered, so that in the event of the trigger being drawn before the bull's eye is reached the target will be hit, and assuming the target to be a man he would be disabled and the object gained. Another important **reason** for advocating the use of the second finger in drawing the trigger **is the fact that the** weight of the military revolver (**2 lbs.** 8 oz.), together with the power required to draw **the** trigger (5lbs. pull), by the long tension of the muscles of the arm, in aiming, causes a vibration, so that the farther the bullet has to travel the farther it is thrown off the centre of the objective. The first finger, therefore, placed along the barrel or side of the pistol, acting as a lever, tends to reduce almost to a minimum the spasmodic muscular vibration; again, in drawing the trigger with the forefinger the hardness of the pull tends to depress the muzzle, while with using the second finger as **before** described this depression is almost impossible.

In rifle-shooting, as also in that of the pistol **and** revolver, the ordinary method should be reversed; that is, instead of commencing at 100 yards from the target, the practice should commence at the longest range, and the target should be gradually approached as if it were an actual enemy.

In revolver practice I would recommend all who desire to become thoroughly efficient to commence at say 100 yards from the target, and to gradually reduce the range to not less than 20 yards. This would accustom the practitioner to get a thorough knowledge of the capabilities of the weapon, and to learn the required amount of elevation necessary. It must be remembered that the Military Regulation Revolver will kill at 300 yards.

I have myself shot with a 320-bore revolver, eight grains of powder, bullet eighty grains, at a regulation target at 200 yards, and have made very fair practice : in fact, the long range is far preferable for practice, as being not only beneficial, but a more exciting pastime than the ordinary range.

To those who do not possess a regulation iron target, I would recommend one similar to that which I have sometimes used. (*Vide* diagram.) This target is made of a simple framework of wood, covered with canvas and layers of paper pasted thereon. It has the double advantage of having the Martini-

Smith target in the centre, and the remaining portion, having the exact size of a man traced thereon, has one other advantage in at once showing the result of the practice. This target can be used over and over again, as, after use, the perforations can be pasted over with small pieces of paper, and when well riddled, it can be re-covered ; and the thicker it becomes the better.

No one should attempt to fire ball-cartridge anywhere but at a proper range. Firing in small back-gardens, against brick or stone walls and trunks of trees, should never be allowed. Bullets will rebound or go off at a tangent, and do serious mischief.

When a bullet once leaves the muzzle of a rifle, pistol, or revolver, by the evolution of gunpowder-gas, there is no dependence upon it as to where it may stop, or what damage it may do, and bullets upon hitting hard ground will ricochet ; therefore, to those who wish to enjoy security at practice, I would advise the selection of ground free from habitation, or where no people are at work—some secluded spot where there is ample range, and, if possible, a natural hill or mound to receive the bullets.

The military revolver will kill at 300 yards, The Snider artillery carbine at 1800 yards, and the Martini-Henry rifle at 3000 yards.* Too much dependence upon the use of the slide of the back-sight for elevation in rifle practice should be deprecated for more than one reason : *e.g.,* assuming that a man has been firing at 300 yards with his back-sight adjusted to that range, and he is suddenly ordered to advance at the double ; if, at the spur of the moment, he neglects to reduce his sight, the result will follow that every shot will go over the enemy. It is simply idle to suppose for one moment that in the heat of action a soldier could afford to fritter away valuable time, or even be allowed to do so, in adjusting back-sights. He would, if he were properly instructed, when within 300 yards place his back-sight level, and rely upon his own skill in judging what elevation he should use.

* Vide *Minor Tactics,* by Lieut.-Colonel Clery, 1883.

It is better to fire low than high. A low shot will usually **ricochet**, particularly upon striking hard ground, greensward, or **a wet** clay soil, and, consequently, will do damage. Very nearly **two** thirds of the bullets in action are **lost** by **going** over the **heads of** the enemy.

In the instruction of men in the use **of the** rifle valuable time is wasted, and too much importance is attached to useless detail. **Let** a man **be** placed before the ordinary regimental target, at an unknown distance, with **the** figure **of a man traced** thereon, assuming the target to be an enemy similarly armed with himself; **let him understand** that he must take his chance of hitting his man **or being hit** himself; and let **him fire at this** target with the back-sight **level, judging** his own distance **and the** necessary elevation required : this calculation (not **a** very difficult one, after **a** little practice) could easily be come to **while** in the act of loading. The result of the first shot would determine the required elevation, and by taking pains, bull's eyes and centres would soon be obtained.

It is submitted that this mode of procedure would create an interest in the practice of the soldier, tending to cause a healthy reaction ; men would take more pains, and try to beat their comrades, as there would be a greater stimulus to do so than by the present system. Men, as it is, go to their practice without the slightest interest therein, and get rid of the ammunition as soon as possible, in order to get off duty. The real reason why we have such excellent shots in the Volunteers is accounted for by the fact that they not only take an interest in the work, but take pains in everything they do, the result being success.

Much significance is attached to the bull's-eye mania. It should be borne in mind that a man is a large object at which to aim ; that so long as he can be crippled there is no necessity to kill. To disable a man so that he can do no more mischief is sufficient.

Any man can make a scale of elevation in his own mind, and, with practice, fire at any range without putting up the sight, and can fire standing. My theory is as follows :—

Up to 100 yards the range is point-blank, that is, aim direct on the bull's eye ; for 200 yards, raise the muzzle, say one foot above the bull's eye ; for 300 yards, two feet above the bull's eye, and so on. A few trial-shots will soon settle the question, and practice makes perfect. A man will thus be independent of the back-sight of his rifle. This refers to shooting in the open. Of course, under cover, when time and circumstances admit, the back-sight can be used with great advantage.

A man in shooting with a pistol or revolver has to judge his own distance and the necessary elevation. Why should not the same rule apply directly to the rifle ? I have seen excellent practice at 400 yards with a Snider carbine, back-sight level, the man judging his own elevation, and have been very successful myself, and have found the above rule apply, with slight variations.

In rifle contests all artificial nonsense, such as coloured glasses, eye-shades, kneeling upon eider-down quilts, firing from shaded tents, blackening sights, &c., should be discouraged.

Let a man leave all such **effeminacy** and tom-**foolery at** home, and shoot like a man, taking circumstances as he would find **them in the** open field *with an enemy* before him, using **such** cover only as nature **and** circumstances provide.

There **is infinite** satisfaction **attached to the** winning of **an** honour, when that honour has **to be** obtained under difficulties which must **be** surmounted. The more difficult the task is, the more merit in overcoming **it**.

Lastly. All firearms **require** constant attention, and should be kept clean. After use they **should be** immediately attended **to, and never put away dirty;** should be kept in some dry **corner where rust** cannot **destroy,** and they **should be occasionally** overhauled and oiled when necessary. **Really** valuable weapons **are** sometimes **ruined** by neglect. The man who takes no pride in his gun is no sportsman.

LONDON:

Printed by STRANGEWAYS AND SONS, Tower Street, Upper St. Martin's Lane